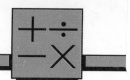

MW00534704

Table Of Contents

Glossary

Angle. Two rays with the same end point.

Area. The number of square units needed to cover a region.

Centimeter. A metric system measurement. There are 2.54 centimeters in·an inch.

Cup (c.). A unit of volume in the customary system equal to 8 ounces.

Decimal. A number with one or more places to the right of a decimal point, such as 6.5 or 2.25.

Denominator. The number below the fraction bar in a fraction.

Diameter. A line segment that passes through the center of a circle and has both end points on the circle.

Digit. The symbols used to write numbers: 0, 1, 2, 3, 4, 5, 6, 7, 8, and 9.

Dividend. The larger number that is divided by the smaller number, or divisor, in a division problem. In the problem 28 ÷ 7 = 4, 28 is the dividend.

Divisor. The number that is divided into the dividend in a division problem. In the problem 28 ÷ 7 = 4, 7 is the divisor.

Equivalent Fractions. Fractions that name the same number.

Estimate. To give an approximate rather than an exact answer.

Factor. The numbers multiplied together in a multiplication problem.

Fraction. A number that names part of a whole, such as 1/2 or 1/3.

Kilometer (km). A unit of length. There are 1000 meters in a kilometer.

Liter (L). A unit in the metric system used to measure amounts of liquid.

Meter (m). A unit of length in the metric system. A meter is equal to 39.37 inches.

Mile (mi.). A mile is equal to 1760 yards.

Mixed Numeral. A number written as a whole number and a fraction.

Multiple. The product of a specific number and any other number. For example, the multiples of 2 are 2 (2 x 1), 4 (2 x 2), 6, 8, 10, 12, and so on.

Numerator. The number above the fraction bar in a fraction.

Octagon. A polygon with eight sides.

Ordered Pair. A pair of numbers used to locate a point in a plane.

Pentagon. A polygon with five sides.

Perimeter. The distance around an object. Found by adding the lengths of the sides.

Pint (pt). A unit of volume in the customary system equal to 2 cups.

Polygon. A closed plane figure with straight sides called line segments.

Product. The answer of a multiplication problem.

Quart (qt). A unit of volume equal to four cups or two pints.

Quotient. The answer of a division problem.

Radius. A line segment with one endpoint on the circle and the other end point at the center.

Rectangle. A figure with four corners and four sides. Sides opposite each other are the same length.

Regroup. To use one ten to form ten ones, one 100 to form ten tens, fifteen ones to form one ten and five ones, and so on.

Remainder. The number left over in the quotient of a division problem.

Rounding. Expressing a number to the nearest ten, hundred, thousand, and so on. For example, round 18 up to 20; round 11 down to 10.

Sequencing. Putting numbers in the correct order, such as 7, 8, 9.

Square. A figure with four corners and four sides of the same length.

Triangle. A figure with three corners and three sides.

Yard. A measurement of distance in the customary system. There are three feet in a yard.

Name: _____

Place Value

Place value is the value of a digit, or numeral, shown by where it is in the number.
For example, in the number 1234, 1 has the place value of thousands, 2 is hundreds, 3 is tens, and 4 is ones.

Directions: Put the numbers in the correct boxes to find how far the car has traveled.

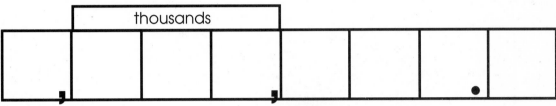

one thousand
six hundreds
eight ones
nine ten thousands
four tens
two millions
seven tenths
five hundred thousands

How many miles has the car traveled? _____

Directions:

In the number:

2386	_____ is in the ones place.
4957	_____ is in the hundreds place.
102,432	_____ is in the ten thousands place.
489,753	_____ is in the one thousands place.
1,743,998	_____ is in the millions place.
9,301,671	_____ is in the hundred thousands place.
7,521,834	_____ is in the tens place.

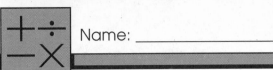

Name: _____

Addition

Addition is "putting together" or adding two or more numbers to find the sum. Regrouping is to use one ten to form ten ones, one 100 to form ten tens, fifteen ones to form one ten and five ones, and so on.

Directions: Add using regrouping. Color in all of the boxes with a 5 in the answer to help the dog find its way home.

	63 +22	5268 4910 +1683	248 +463	291 +543	2934 +112
1736 +5367	2946 +7384	3245 1239 +981	738 +692	896 +729	594 +738
2603 +5004	4507 +289	1483 +6753	1258 +6301	27 469 +6002	4637 +7531
782 +65	485 +276	3421 +8064			
48 93 +26	90 263 +864	362 453 +800			

Name: _____

Subtraction

Subtraction is "taking away" or subtracting one number from another. Regrouping is to use one ten to form ten ones, one 100 to form ten tens, fifteen ones to form one ten and five ones, and so on.

Directions: Subtract using regrouping.

Examples:

```
  23
 -18
   5
```

```
 243
 -96
 147
```

81	76	94	156	243	468
-53	-49	-38	-77	-29	-293
341	568	806	647	730	961
-83	-173	-738	-289	-518	-846
573	604	254	111	358	147
-76	-55	-69	-82	-99	-49
265	372	180	325	873	726
-19	-59	-106	-68	-35	-29

Name: _____

Rounding

Rounding a number means expressing it to the nearest ten, hundred, thousand, and so on.

Directions: Round the following numbers to the nearest ten. If the number has 5 ones or more, round it up to the next highest ten. For example, round 26 up to 30. If the number has 4 ones or less, round down to the nearest ten, such as rounding 44 down to 40.

18 _____   33 _____   82 _____   56 _____

24 _____   49 _____   91 _____   67 _____

Directions: Round to the nearest hundred. If 5 tens or more, round up. If 4 tens or less, round down.

243 _____   689 _____   263 _____   162 _____

389 _____   720 _____   351 _____   490 _____

463 _____   846 _____   928 _____   733 _____

Directions: Round to the nearest thousand. If number has 5 hundreds or more, round up. If 4 hundreds or less, round down.

2638 _____   3940 _____   8653 _____   6238 _____

1429 _____   5061 _____   7289 _____   2742 _____

9460 _____   3109 _____   4697 _____   8302 _____

Directions: Round to the nearest ten thousand. If the number has 5 thousands or more, round up. If 4 thousands or less, round down.

11,368 _____   38,421 _____   75,302 _____   67,932 _____

14,569 _____   49,926 _____   93,694 _____   81,648 _____

26,784 _____   87,065 _____   57,843 _____   29,399 _____

Name: _____

Addition And Subtraction

Addition is "putting together" or adding two or more numbers to find the sum. Subtraction is "taking away" or subtracting one number from another.

Regrouping is to use one ten to form ten ones, one 100 to form ten tens, fifteen ones to form one ten and five ones, and so on.

Directions: Add or subtract. Remember to regroup.

```
  32        183        456        643
  68        246        398       -377
 +43        +89       +597
```

```
 1563       3586       8711       9361       5734
 -941      +4218      -4937      -7452      +6298
```

```
  293        743        849       1227       9117
  431       -529        250       2431      -3828
  +93                   +82      +5792
```

68 + 93 + 146 = _____   73 + 246 + 1579 = _____

43 + 745 - 29 = _____   128 + 403 + 2571= _____

156 + 627 + 541 = _____   97 + 51 + 37 + 79 = _____

Tom walks 389 steps from his house to the video store. It is 149 steps to Elm Street. It is 52 steps from Maple Street to the video store. How many steps is it from Elm Street to Maple Steet?

_____

Name: _____

Addition And Subtraction

Addition is "putting together" or adding two or more numbers to find the sum. Subtraction is "taking away" or subtracting one number from another.

Directions: Add or subtract.

```
    38        1269       5792        629        4697
    43        2453      -4814        491       -2988
   +21       +8219                  +308


  5280          68        197       7321         456
 -3147          27        436      -2789        +974
               +42       +213


  3932         492       9873       4978        6235
 +4681         863      +5483      +2131       +2986
               +57
```

Sue stocked her pond with 263 bass and 187 trout. The turtles ate 97 fish.
How many fish are left?   _____

Multiples

A multiple is the product of a specific number and any other number. For example, the multiples of 2 are 2 (2 x 1) 4, (2 x 2), 6, 8, 10, 12, and so on.

Directions: Write the missing multiples.

Example: Count by fives.
5, 10, 15, 20, 25, 30, 35. These are multiples of 5.

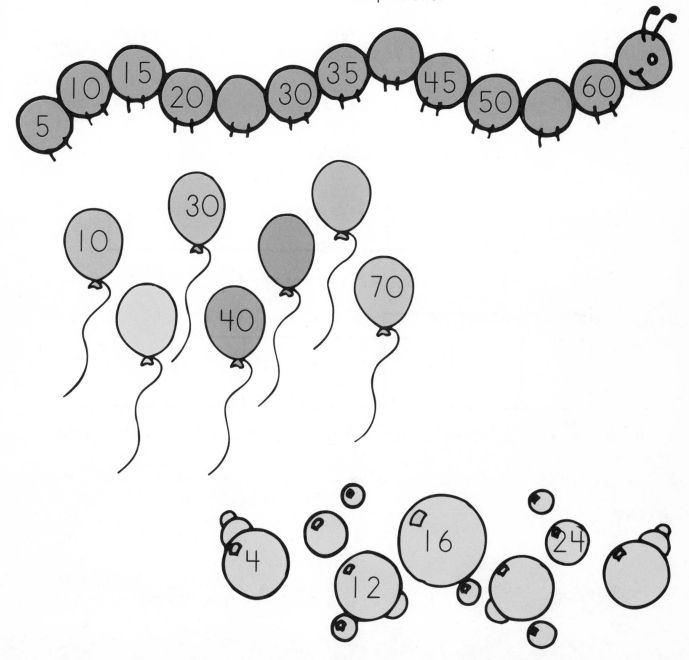

 ÷
+   ÷
− ×   Name: _____

Review

Directions: Add or subtract using regrouping.

67	5029	732	2467	8453
93	-3068	801	+3184	-6087
+48		+18		

5792	7489	463	3537	6342
-3889	+5938	-209	-2394	+959

Directions: Write the numbers in the boxes.

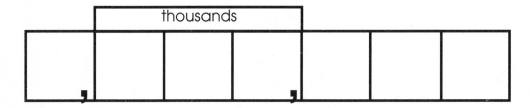

eight million, four hundred thousand, nine hundred fifty two

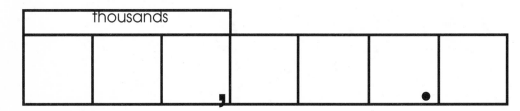

five hundred thousands, three ten thousands, five thousands, zero hundreds, four tens, one one, two tenths

Directions: Fill in the blanks with the missing multiples.

6, 12, 18, _____, 30, _____   3, _____, _____, 12, 15

4, _____, 12, 16, _____, 24   _____, 10, 15, _____, _____

Multiplication

Multiplication is a short way to find the sum of adding the same number a certain amount of times, such as 7 x 4 = 28 instead of 7 + 7 + 7 + 7 = 28.

Directions: Multiply as fast as you can.

4 x7	7 x6	0 x8		
7 x2	9 x5	1 x5		6 x4
8 x3	7 x1	4 x2		9 x6
8 x5	6 x7	9 x8	3 x5	7 x8
3 x9	5 x6	9 x9	7 x5	9 x4
3 x6	2 x8		8 x6	7 x7
0 x7			3 x3	5 x9

Name: _____

Multiplication: Tens, Hundreds, And Thousands

Multiplication is a short way to find the sum of adding the same number a certain amount of times, such as 7 x 4 = 28 instead of 7 + 7 + 7 + 7 = 28.

Directions: Study the examples.

Examples:

When multiplying a number by 10, the answer is the number with a zero.
It is like counting by 10s.

10	10	10	10	10	10
x1	x2	x3	x4	x5	x6
10	20	30	40	50	60

When multiplying a number by 100, the answer is the number with two zeroes.
When multiplying a number by 1000, the answer is the number with three zeroes.

100	100	100	1000	1000	1000
x1	x2	x3	x1	x2	x3
100	200	300	1000	2000	3000

Such basic facts help us multiply.

4	400	8	800	7	700
x2	x2	x3	x3	x5	x5
8	800	24	2400	35	3,500

Directions: Multiply.

10	60	400	700	50
x3	x5	x5	x8	x7

80	4000	6000	300	700
x9	x2	x4	x9	x6

3 x 800 = _____   9 x 2000 = _____   7 x 90 = _____

Name: _____

Multiplication: One-Digit Number x Two-Digit Number

Multiplication is a short way to find the sum of adding the same number a certain amount of times, such as 7 x 4 = 28 instead of 7 + 7 + 7 + 7 = 28.

Directions: Study the example. Follow the steps to multiplying by regrouping tens.

Example:

Step 1. Multiply ones. Regroup.

$$\begin{array}{r} {}^{2}54 \\ \underline{\times 7} \\ 8 \end{array}$$

Step 2. Multiply Tens. Add 2 tens.

$$\begin{array}{r} {}^{2}54 \\ \underline{\times 7} \\ 378 \end{array}$$

27 x3	63 x4	52 x5	91 x9	45 x7
75 x2	64 x5	76 x3	93 x6	87 x4
66 x7	38 x2	47 x8	64 x9	51 x8
99 x3	13 x7	32 x4	25 x8	15 x7

The chickens on the Smith farm produce 48 dozen eggs each day. How many dozen eggs do they produce in 7 days? _____

Name: _____

Multiplication: Two-Digit Number x Two-Digit Number

Multiplication is a short way to find the sum of adding the same number a certain amount of times, such as 7 x 4 = 28 instead of 7 + 7 + 7 + 7 = 28.

Directions: Study the examples. Follow the steps to multiply by regrouping.

Example:

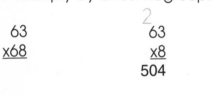

Step 1. Multiply by ones. Regroup.   Step 2. Multiply by tens. Regroup. Add.

```
              2
  63         63          63          63
 x68         x8         x60         x68
            504        3780         504
                                   3780
                                   4284
```

```
  12          27          65          19
 x55         x15         x27         x39
```

```
  99          35          43          38
 x13         x14         x26         x17
```

```
  53          47          57          48
 x86         x72         x62         x33
```

```
  27          93          64          53
 x54         x45         x16         x23
```

The Jones farm has 24 cows that each produce 52 quarts of milk a day. How many quarts are produced each day altogether? _____

Name: _____

Multiplication: Two-Digit Number x Three-Digit Number

Multiplication is a short way to find the sum of adding the same number a certain amount of times, such as 7 x 4 = 28 instead of 7 + 7 + 7 + 7 = 28.

Directions: Study the example. Follow the steps to multiply.

Example:

Step 1. Multiply by ones. Regroup.   Step 2. Multiply by tens. Regroup. Add.

```
                  2 2
   287        287                      287        287
   x43         x3                      x40        x43
              861                   11,480        861
                                              11,480
                                              12,341
```

```
   261        434        357          614        368
   x36        x48        x75          x59        x98
```

```
   231        754        549          372        458
   x46        x65        x89          x94        x85
```

At the Douglas berry farm, workers pick 378 baskets of strawberries each day. Each basket holds 65 strawberries. How many strawberries are picked each day?

_____

Name: _____

Multiplication: Three-Digit Number x Three-Digit Number

Multiplication is a short way to find the sum of adding the same number a certain amount of times, such as 7 x 4 = 28 instead of 7 + 7 + 7 + 7 = 28.

Directions: Multiply. Regroup when needed.

Example:

```
     563
    x248
    4504
   22520
  112600
  139,624
```

Hint: When multiplying by the tens, start writing the number in the tens place. When multiplying by the hundreds, start in the hundreds place.

```
   842        932        759        531
  x167       x272       x468       x556
```

```
   383        523        229        738
  x476       x349       x189       x513
```

```
   483        946        365
  x148       x367       x622
```

James grows pumpkins on his farm. He has 362 rows of pumpkins. There are 593 pumpkins in each row. How many pumpkins does James grow? _____

Multiplication

Multiplication is a short way to find the sum of adding the same number a certain amount of times, such as 7 x 4 = 28 instead of 7 + 7 + 7 + 7 = 28.

Directions: Multiply. Use your answers to follow the code to color the quilt.

70,725 — red	448 — white	34,088 — blue
667 — green	249,738 — orange	221,446 — yellow

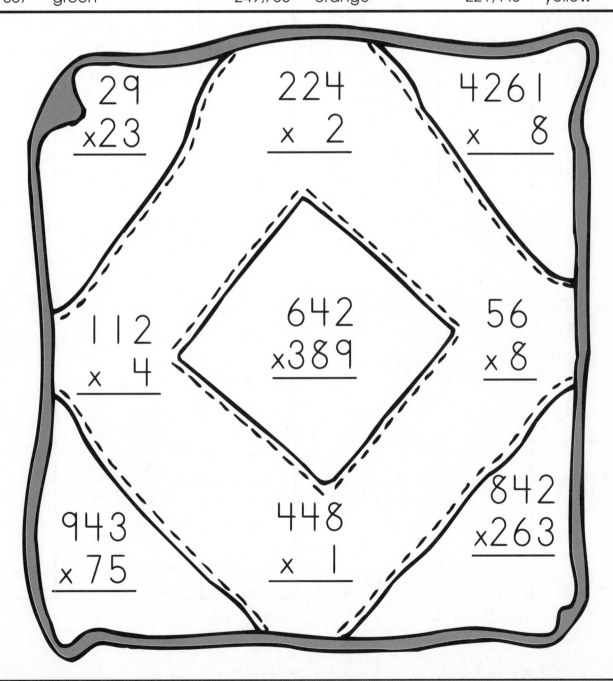

Name: _____

Review

Directions: Multiply. Work the problem in the box. Color the ribbons blue if the answer is correct.

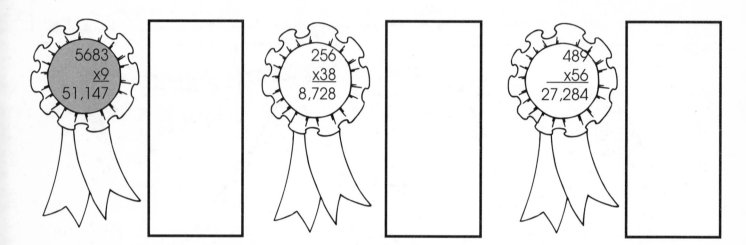

5683
x9
51,147

256
x38
8,728

489
x56
27,284

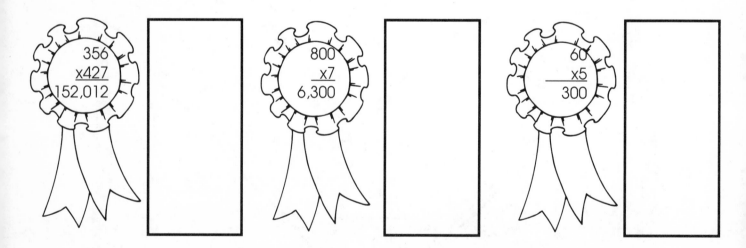

356
x427
152,012

800
x7
6,300

60
x5
300

Name: _____

Division

Division is a way to find out how many times one number is contained in another number. For example, 28 ÷ 7 = 4 means that there are four groups of seven in 28.

Directions: Study the example. Then divide to solve the problems. Remember that the remainder must be smaller than the divisor.

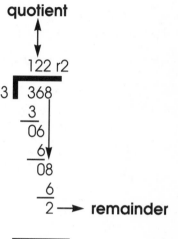

divisor

$$3\overline{)368}$$

dividend

$$3\overline{)368} \\ \underline{3} \\ 06$$
(quotient: 1)

$$3\overline{)368} \\ \underline{3} \\ 06 \\ \underline{6} \\ 08$$
(quotient: 12)

quotient

$$3\overline{)368} \\ \underline{3} \\ 06 \\ \underline{6} \\ 08 \\ \underline{6} \\ 2$$
(quotient: 122 r2) → **remainder**

$$7\overline{)860} \qquad 6\overline{)611} \qquad 8\overline{)279} \qquad 4\overline{)338} \qquad 6\overline{)979}$$

$$3\overline{)792} \qquad 5\overline{)463} \qquad 6\overline{)940} \qquad 4\overline{)647} \qquad 3\overline{)814}$$

$$7\overline{)758} \qquad 5\overline{)356} \qquad 4\overline{)276} \qquad 8\overline{)328} \qquad 9\overline{)306}$$

The record store has 491 records. The store sells 8 records a day. How many days will it take to sell all of the records? _____

Name: _____

Division: Checking The Answer

Division is a way to find out how many times one number is contained in another number. To check a division problem, multiply the quotient by the divisor. Add the remainder. The answer will be the dividend.

Directions: Study the example. Divide to work the problems. Draw a line from the division problem to the correct checking problem.

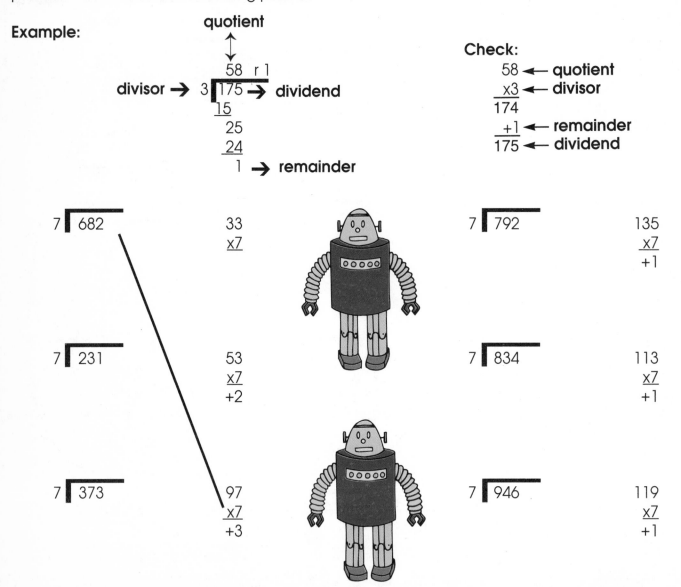

The toy factory puts 7 robots in each box. The factory has 256 robots. How many boxes will they need? _____

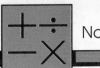

Name: _____

Division: 1-Digit Divisor

Division is a way to find out how many times one number is contained in another number.

Directions: Work the problems on another sheet of paper. Use the code to color the picture.

Color these answers:				
5 ⟌ 895	6 ⟌ 493	6 ⟌ 940	4 ⟌ 647	**orange**
4 ⟌ 672	6 ⟌ 696	5 ⟌ 749	8 ⟌ 628	**blue**
3 ⟌ 814	7 ⟌ 490	5 ⟌ 398	2 ⟌ 571	**black**

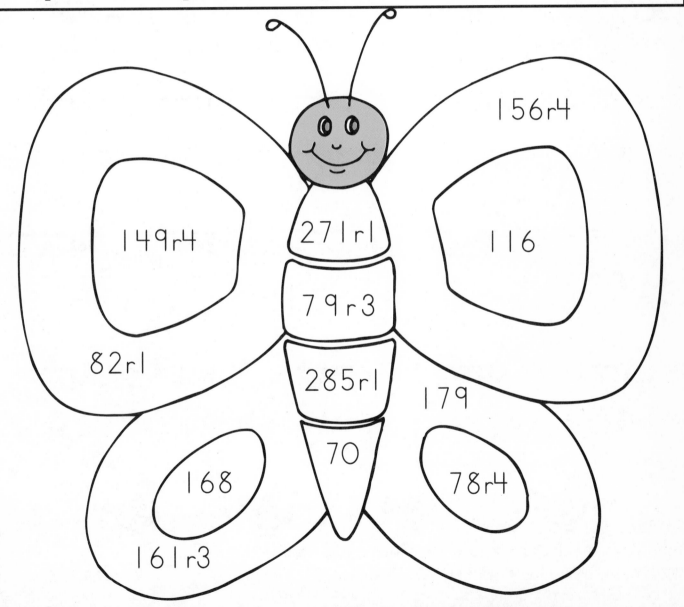

Division: 2-Digit Divisor

Division is a way to find out how many times one number is contained in another number.

Directions: Study the example. Divide. Remember to check your answer by multiplying it by the divisor and adding the remainder.

Example:

```
        2
   12 | 256
       24
        1
```

```
       21  r4
   12 | 256
       24↕
        16
        12
         4
```

```
Check:   21
        x12
         42
         21
        252
         +4
        256
```

27 | 880   81 | 913   65 | 790   42 | 674   67 | 823

72 | 977   54 | 743   45 | 863   24 | 432   18 | 372

28 | 175   49 | 538   77 | 936   37 | 603   63 | 835

The Allen farm has 882 chickens. The chickens are kept in 21 coops.
How many chickens are there in each coop?   _____

Name: _____

Division: Checking The Answer

Division is a way to find out how many times one number is contained in another number.

Directions: Divide, then check your answers.

Example:

```
    182  r1        Check:    182
4 ⎡729                        x4
    4                        728
    32                        +1
    32                       729
     9
     8
     1
```

```
35 ⎡468        check:   ⎡_____⎤
                        ⎣  x35 ⎦
```

```
77 ⎡819        check:   ⎡_____⎤
                        ⎣  x77 ⎦
```

```
29 ⎡568        check:   ⎡_____⎤
                        ⎣  x29 ⎦
```

```
53 ⎡2795       check:   ⎡_____⎤
                        ⎣  x53 ⎦
```

```
67 ⎡2856       check:   ⎡_____⎤
                        ⎣  x67 ⎦
```

```
41 ⎡6382       check:   ⎡_____⎤
                        ⎣  x41 ⎦
```

The bookstore puts 53 books on a shelf. How many shelves will it need for 1590 books? _____

Name: _____

Review

Directions: Divide.

3 | 268   15 | 165   27 | 489   48 | 695

79 | 937   49 | 683   91 | 848   73 | 592   59 | 473

23 | 1268   67 | 2543   81 | 3608   37 | 8432   97 | 4528

Directions: Find the averages. An average is found by adding two or more quantities and then dividing by the number of quantities.

22, 38 _____   105, 263, 331 _____

48, 100, 62 _____   248, 325, 250, 69 _____

17, 18, 36, 28, 6 _____   87, 91, 55, 48, 119 _____

Name: _____

Fraction: Addition

A fraction is a number that names part of a whole, such as 1/2 or 1/3. The denominator is the bottom number in a fraction; the numerator is the top number.

When adding fractions with the same denominator, the denominator stays the same. Add only the numerators.

Example:

numerator → $\dfrac{1}{8}$ + $\dfrac{2}{8}$ = $\dfrac{3}{8}$
denominator →

Directions: Study the example. Add the fractions. The first one is done for you.

Name: _____

Fractions: Subtraction

A fraction is a number that names part of a whole, such as 1/2 or 1/3. The denominator is the bottom number in a fraction; the numerator is the top number.

When subtracting fractions with the same denominator, the denominator stays the same. Subtract only the numerators.

Directions: Solve the problems below, working from left to right across each row. As you find each answer, copy the letter from the code box into the numbered blanks. The first one is done for you. The answer will tell the name of a famous American.

1. $\dfrac{3}{8} - \dfrac{2}{8} = \dfrac{1}{8}$

2. $\dfrac{2}{4} - \dfrac{1}{4} = $ _____

3. $\dfrac{5}{9} - \dfrac{3}{9} = $ _____

4. $\dfrac{2}{3} - \dfrac{1}{3} = $ _____

5. $\dfrac{8}{12} - \dfrac{7}{12} = $ _____

6. $\dfrac{4}{5} - \dfrac{1}{5} = $ _____

7. $\dfrac{6}{12} - \dfrac{3}{12} = $ _____

8. $\dfrac{4}{9} - \dfrac{1}{9} = $ _____

9. $\dfrac{11}{12} - \dfrac{7}{12} = $ _____

10. $\dfrac{7}{8} - \dfrac{3}{8} = $ _____

11. $\dfrac{4}{7} - \dfrac{2}{7} = $ _____

12. $\dfrac{14}{16} - \dfrac{7}{16} = $ _____

13. $\dfrac{18}{20} - \dfrac{13}{20} = $ _____

14. $\dfrac{13}{15} - \dfrac{2}{15} = $ _____

15. $\dfrac{5}{6} - \dfrac{3}{6} = $ _____

Code Box				
T 1/8	p 5/24	h 1/4	f 4/12	e 2/7
J 3/12	e 3/9	o 2/9	f 4/8	r 7/16
o 2/8	y 8/20	q 1/32	m 1/3	s 5/20
a 1/12	r 12/15	s 3/5	n 2/6	o 11/15

Who helped write the Declaration of Independence?

1. _____ 2. _____ 3. _____ 4. _____ 5. _____ 6. _____

7. _____ 8. _____ 9. _____ 10. _____ 11. _____ 12. _____ 13. _____ 14. _____ 15. _____

Math

÷
+
− ×

Name: _____

Fractions: Adding Mixed Numerals

A mixed numeral is a number written as a whole number and a fraction, such as 6 5/8.

Directions: Add the number in the center to the numbers in the rings.

Example:

$$9 \frac{1}{3}$$
$$+3 \frac{1}{3}$$
$$\overline{12 \frac{2}{3}}$$

$$2 \frac{3}{6}$$
$$+1 \frac{1}{6}$$
$$\overline{3 \frac{4}{6}}$$

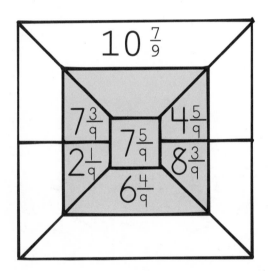

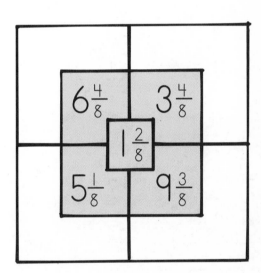

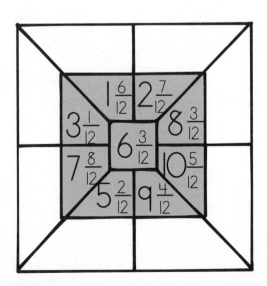

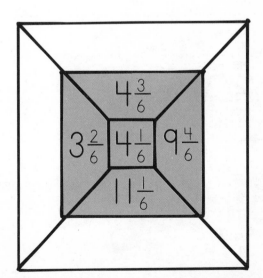

Name: _____

Fractions: Subtracting Mixed Numerals

A mixed numeral is a number written as a whole number and a fraction, such as 6 5/8.

Directions: Solve the problems. The first one is done for you.

$7 \frac{3}{8}$
$-4 \frac{2}{8}$
$3 \frac{1}{8}$

$4 \frac{5}{6}$
$-3 \frac{1}{6}$

$4 \frac{1}{2}$
-3

$7 \frac{5}{8}$
$-6 \frac{3}{8}$

$6 \frac{6}{8}$
$-1 \frac{1}{8}$

$5 \frac{3}{4}$
$-1 \frac{1}{4}$

$5 \frac{2}{3}$
$-3 \frac{1}{3}$

$4 \frac{8}{10}$
$-3 \frac{3}{10}$

$9 \frac{8}{9}$
$-4 \frac{3}{9}$

$7 \frac{2}{3}$
$-6 \frac{1}{3}$

$7 \frac{2}{3}$
-5

$9 \frac{8}{10}$
$-6 \frac{3}{10}$

$4 \frac{7}{9}$
-2

$6 \frac{7}{8}$
$-5 \frac{3}{8}$

$6 \frac{3}{4}$
$-3 \frac{1}{4}$

$5 \frac{6}{7}$
$-3 \frac{1}{7}$

$7 \frac{6}{7}$
$-2 \frac{4}{7}$

Sally needs 1 3/8 yards of cloth to make a dress. She has 4 5/8 yards. How much will be left over?

_____

Name: _____

Fractions: Equivalent

Equivalent fractions name the same number, such as 1/2 and 2/4.

Directions: Study the example. Draw a line between the equivalent fractions in each row.

Example:

Equivalent fractions are equal to each other.

$$1/2 = 2/4 = 4/8$$

$\frac{1}{2}$	$\frac{6}{16}$	$\frac{8}{12}$	$\frac{8}{56}$
$\frac{2}{3}$	$\frac{2}{4}$	$\frac{12}{32}$	$\frac{8}{16}$
$\frac{3}{4}$	$\frac{6}{12}$	$\frac{12}{16}$	$\frac{24}{32}$
$\frac{2}{5}$	$\frac{6}{8}$	$\frac{4}{8}$	$\frac{24}{64}$
$\frac{3}{6}$	$\frac{4}{6}$	$\frac{12}{24}$	$\frac{16}{24}$
$\frac{3}{8}$	$\frac{2}{6}$	$\frac{8}{20}$	$\frac{8}{24}$
$\frac{1}{9}$	$\frac{2}{14}$	$\frac{4}{36}$	$\frac{24}{96}$
$\frac{1}{3}$	$\frac{4}{10}$	$\frac{4}{12}$	$\frac{16}{40}$
$\frac{1}{7}$	$\frac{6}{24}$	$\frac{4}{28}$	$\frac{24}{48}$
$\frac{3}{12}$	$\frac{2}{18}$	$\frac{12}{48}$	$\frac{8}{72}$

Name: _____

Fractions: Reducing

Reducing a fraction means to find the greatest common factor and divide.

Directions: Reduce each fraction. Circle the answer.

Example: $\dfrac{5}{15} = \dfrac{1}{3}$ factors of 5:   1, 5 $5 \div 5 = 1$
                                   factors of 15:   1, 3, 5, 15 $15 \div 5 = 3$

$\dfrac{2}{4} = \dfrac{1}{2}, \dfrac{1}{6}, \dfrac{1}{8}$ $\dfrac{3}{9} = \dfrac{1}{6}, \dfrac{1}{3}, \dfrac{3}{6}$ $\dfrac{5}{10} = \dfrac{1}{5}, \dfrac{1}{2}, \dfrac{5}{6}$

$\dfrac{4}{12} = \dfrac{1}{4}, \dfrac{1}{3}, \dfrac{2}{3}$ $\dfrac{10}{15} = \dfrac{2}{3}, \dfrac{2}{5}, \dfrac{2}{7}$ $\dfrac{12}{14} = \dfrac{1}{8}, \dfrac{6}{7}, \dfrac{3}{5}$

$\dfrac{3}{24} = \dfrac{2}{12}, \dfrac{3}{6}, \dfrac{1}{8}$ $\dfrac{1}{11} = \dfrac{1}{11}, \dfrac{2}{5}, \dfrac{3}{4}$ $\dfrac{11}{22} = \dfrac{1}{12}, \dfrac{1}{2}, \dfrac{2}{5}$

Directions: Find the way home. Color the boxes with fractions equivalent to 1/8 and 1/3.

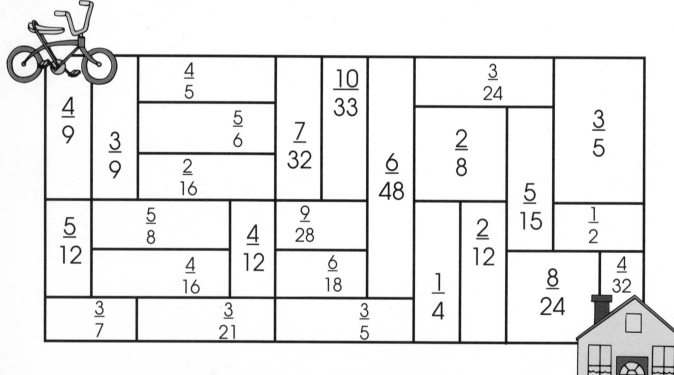

Name: _____

Fractions: Mixed Numerals

A mixed numeral is a number written as a whole number and a fraction, such as 6 5/8.

Directions: Change each fraction to a mixed numeral. Make the mixed numerals into fractions.

Example:

To change a fraction into a mixed numeral, divide the denominator (bottom number) into the numerator (top number). Put the remainder over the denominator.

To change a mixed numeral into a fraction, multiply the denominator by the whole number, add the numerator, and place it on top of the denominator.

$\frac{14}{6}$ = $2\frac{2}{6}$

$3\frac{1}{7}$ = $\frac{22}{7}$ (7 × 3) + 1 = $\frac{22}{7}$

$\frac{21}{6}$ = ____

$\frac{24}{5}$ = ____

$\frac{10}{3}$ = ____

$\frac{21}{4}$ = ____

$\frac{11}{6}$ = ____

$\frac{13}{4}$ = ____

$\frac{12}{5}$ = ____

$\frac{10}{9}$ = ____

$4\frac{3}{8}$ = $\frac{\square}{8}$

$2\frac{1}{3}$ = $\frac{\square}{3}$

$4\frac{3}{5}$ = $\frac{\square}{5}$

$3\frac{4}{6}$ = $\frac{\square}{6}$

$7\frac{1}{4}$ = $\frac{\square}{4}$

$2\frac{3}{5}$ = $\frac{\square}{5}$

$7\frac{1}{2}$ = $\frac{\square}{2}$

$6\frac{5}{7}$ = $\frac{\square}{7}$

$\frac{11}{8}$ = ____

$\frac{21}{4}$ = ____

$\frac{33}{5}$ = ____

$\frac{13}{6}$ = ____

$\frac{23}{7}$ = ____

$8\frac{1}{3}$ = ____

$9\frac{3}{7}$ = ____

$\frac{32}{24}$ = ____

Name: _____

Review

Directions: Add or subtract the fractions and mixed numerals.

$\frac{3}{8} - \frac{1}{8} =$ ___ $\frac{3}{4} - \frac{2}{4} =$ ___ $\frac{3}{5} + \frac{1}{5} =$ ___ $\frac{4}{12} + \frac{3}{12} =$ ___ $\frac{3}{9} + \frac{1}{9} =$ ___

$$3 \; \frac{1}{8}$$
$$+1 \; \frac{3}{8}$$

$$4 \; \frac{5}{6}$$
$$-3 \; \frac{1}{6}$$

$$7 \; \frac{5}{11}$$
$$+3 \; \frac{3}{11}$$

$$8 \; \frac{3}{9}$$
$$+2 \; \frac{5}{9}$$

$$4 \; \frac{7}{8}$$
$$-2 \; \frac{5}{8}$$

Directions: Reduce the fractions. Circle the answers.

$\frac{3}{6} =$ $\frac{1}{7}$ $\frac{1}{2}$ $\frac{1}{4}$	$\frac{2}{8} =$ $\frac{1}{3}$ $\frac{1}{4}$ $\frac{1}{16}$	$\frac{4}{6} =$ $\frac{1}{4}$ $\frac{2}{3}$ $\frac{3}{9}$
$\frac{4}{20} =$ $\frac{1}{4}$ $\frac{1}{3}$ $\frac{1}{5}$	$\frac{7}{21} =$ $\frac{1}{7}$ $\frac{1}{3}$ $\frac{1}{5}$	$\frac{9}{12} =$ $\frac{3}{5}$ $\frac{1}{8}$ $\frac{3}{4}$

Directions: Reduce the fractions.

$\frac{6}{24} =$ ____ $\frac{8}{32} =$ ____ $\frac{2}{4} =$ ____

$\frac{3}{15} =$ ____ $\frac{6}{12} =$ ____ $\frac{3}{9} =$ ____

Directions: Change the mixed numerals to fractions and the fractions to mixed numerals.

$3 \frac{1}{3} = \frac{\Box}{3}$ $\frac{14}{4} =$ ___ $\frac{26}{6} =$ ___ $3 \frac{7}{12} = \frac{\Box}{12}$ $\frac{22}{7} =$ ___

ANSWER KEY

*This Answer Key has been designed so that
it may be easily removed if you so desire.*

MASTER MATH
4

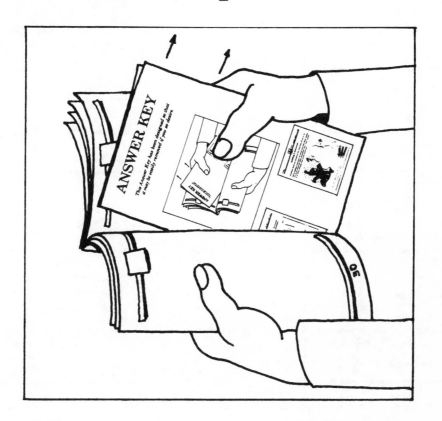

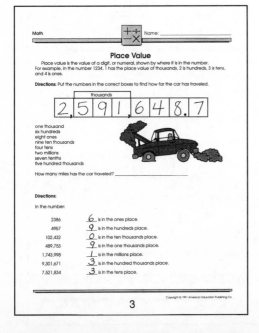

Math   Name: _____

Place Value

Place value is the value of a digit, or numeral, shown by where it is in the number.
For example, in the number 1234, 1 has the place value of thousands, 2 is hundreds, 3 is tens,
and 4 is ones.

Directions: Put the numbers in the correct boxes to find how far the car has traveled.

		thousands					
2,	5	9	1,	6	4	8.	7

one thousand
six hundreds
eight ones
nine ten thousands
four tens
two millions
seven tenths
five hundred thousands

How many miles has the car traveled? _____

Directions:

In the number:

2386	__6__ is in the ones place.
4957	__9__ is in the hundreds place.
102,432	__0__ is in the ten thousands place.
489,753	__9__ is in the one thousands place.
1,743,998	__1__ is in the millions place.
9,301,671	__3__ is in the hundred thousands place.
7,521,834	__3__ is in the tens place.

3

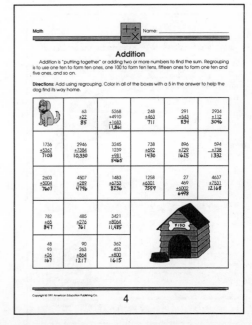

Math   Name: _____

Addition

Addition is "putting together" or adding two or more numbers to find the sum. Regrouping
is to use one ten to form ten ones, one 100 to form ten tens, fifteen ones to form one ten and
five ones, and so on.

Directions: Add using regrouping. Color in all of the boxes with a 5 in the answer to help the
dog find its way home.

🐕	63 +22 **85**	5268 +4910 +1683 **11,861**	248 +463 **711**	291 +543 **834**	2934 +112 **3046**
1736 +5367 **7103**	2946 +7384 **10,330**	3245 1239 +981 **5465**	738 +692 **1430**	896 +729 **1625**	594 +738 **1332**
2603 +5004 **7607**	4507 +289 **4796**	1483 +6753 **8236**	1258 +6301 **7559**	27 469 +6002 **6498**	4637 +7531 **12,168**
782 +65 **847**	485 +276 **761**	3421 +8064 **11,485**			
48 93 +26 **167**	90 263 +864 **1217**	362 453 +800 **1615**			

4

Subtraction

Subtraction is "taking away" or subtracting one number from another. Regrouping is to use one ten to form ten ones, one 100 to form ten tens, fifteen ones to form one ten and five ones, and so on.

Directions: Subtract using regrouping.

Examples:

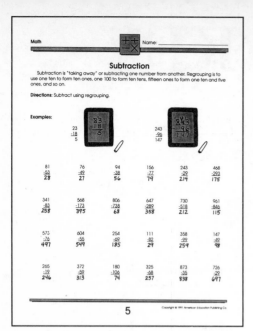

```
 23        243
-18        -96
  5        147
```

```
 81      76      94     156     243     468
-53     -49     -38     -77     -29    -293
 28      27      56      79     214     175

341     568     806     647     730     961
-83    -173    -738    -289    -518    -846
258     395      68     358     212     115

573     604     254     111     358     147
-76     -55     -69     -82     -99     -49
497     549     185      29     259      98

265     372     180     325     873     726
-19     -59    -106     -68     -35     -29
246     313      74     257     838     697
```

Rounding

Rounding a number means expressing it to the nearest ten, hundred, thousand, and so on.

Directions: Round the following numbers to the nearest ten. If the number has 5 ones or more, round it up to the next highest ten. For example, round 26 up to 30. If the number has 4 ones or less, round down to the nearest ten, such as rounding 44 down to 40.

```
18  20     33  30     82  80     56  60
24  20     49  50     91  90     67  70
```

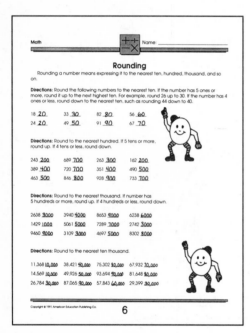

Directions: Round to the nearest hundred. If 5 tens or more, round up. If 4 tens or less, round down.

```
243  200    689  700    263  300    162  200
389  400    720  700    351  400    490  500
463  500    846  800    928  900    733  700
```

Directions: Round to the nearest thousand. If number has 5 hundreds or more, round up. If 4 hundreds or less, round down.

```
2638  3000    3940  4000    8653  9000    6238  6000
1429  1000    5061  5000    7289  7000    2742  3000
9460  9000    3109  3000    4697  5000    8302  8000
```

Directions: Round to the nearest ten thousand.

```
11,368  10,000    38,421  40,000    75,302  80,000    67,932  70,000
14,569  10,000    49,926  50,000    93,694  90,000    81,648  80,000
26,784  30,000    87,065  90,000    57,843  60,000    29,399  30,000
```

Addition And Subtraction

Addition is "putting together" or adding two or more numbers to find the sum. Subtraction is "taking away" or subtracting one number from another.

Regrouping is to use one ten to form ten ones, one 100 to form ten tens, fifteen ones to form one ten and five ones, and so on.

Directions: Add or subtract. Remember to regroup.

```
 32     183     456     643
 68     246     398    -377
+43     +89    +597     264
143     518    1451

1563    3586    8711    9361    5734
-941   +4218   -4937   -7452   +6298
 622    7804    3774    1909   12,032

293     743     849    1227    9117
431     529     250    2431   -3828
+93     214     +82   +5792    5289
817            1181    9450
```

68 + 93 + 146 = **307**

43 + 745 - 29 = **759**

156 + 627 + 541 = **1324**

73 + 246 + 1579 = **1898**

128 + 403 + 2571 = **3102**

97 + 51 + 37 + 79 = **264**

Tom walks 389 steps from his house to the video store. It is 149 steps to Elm Street. It is 52 steps from Maple Street to the video store. How many steps is it from Elm Street to Maple Street? **188**

Addition And Subtraction

Addition is "putting together" or adding two or more numbers to find the sum. Subtraction is "taking away" or subtracting one number from another.

Directions: Add or subtract.

```
 38     1269    5792     629    4697
 43     2453   -4814     491   -2988
+21    +8219     978    +308    1709
102    11,941           1428

5280      68     197    7321     456
-3147     27     436   -2789    +974
2133     +42    +213    4532    1430
         137     846

3932     492    9873    4978    6235
+4681    863   +5483   +2131   +2986
8613     +57   15,356   7109    9221
        1412
```

Sue stocked her pond with 263 bass and 187 trout. The turtles ate 97 fish. How many fish are left? **353**

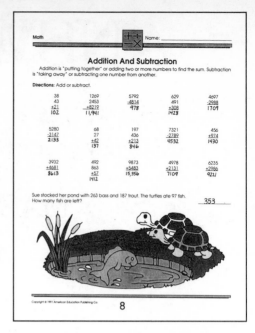

Multiples

A multiple is the product of a specific number and any other number. For example, the multiples of 2 are 2 (2 x 1) 4, (2 x 2), 6, 8, 10, 12, and so on.

Directions: Write the missing multiples.

Example: Count by fives.
5, 10, 15, 20, 25, 30, 35. These are multiples of 5.

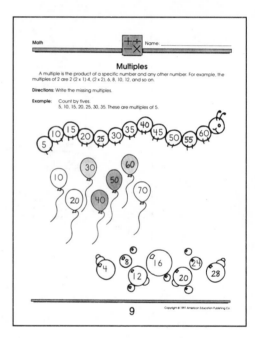

Review

Directions: Add or subtract using regrouping.

```
 67     5029     732    2467    8453
 93    -3068     801   +3184   -6087
+48    1961     +18    5651    2366
208             1551

5792    7489     463    3537    6342
-3889   +5938   -209   -2394    +959
1903    13,427   254    1143    7301
```

Directions: Write the numbers in the boxes.

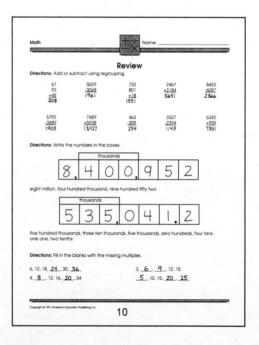

	thousands						
8,	4	0	0,	9	5	2	

eight million, four hundred thousand, nine hundred fifty two

	thousands						
5	3	5,	0	4	1 .	2	

five hundred thousands, three ten thousands, five thousands, zero hundreds, four tens, one one, two tenths

Directions: Fill in the blanks with the missing multiples.

6, 12, 18, **24**, 30, **36**

3, **6**, **9**, 12, 15

4, **8**, 12, 16, **20**, 24

5, 10, 15, **20**, **25**

Multiplication

Multiplication is a short way to find the sum of adding the same number a certain amount of times, such as 7 x 4 = 28 instead of 7 + 7 + 7 + 7 = 28.

Directions: Multiply as fast as you can.

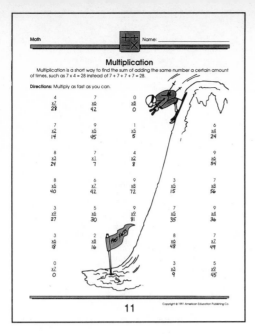

4 x7 = 28	x6 = 42	0 x8 = 0	
7 x2 = 14	9 x5 = 45	1 x5 = 5	6 x4 = 24
8 x3 = 24	7 x1 = 7	4 x2 = 8	9 x6 = 54
8 x5 = 40	6 x7 = 42	9 x8 = 72	3 x5 = 15 7 x8 = 56
3 x9 = 27	5 x6 = 30	9 x9 = 81	7 x5 = 35 9 x4 = 36
3 x6 = 18	2 x8 = 16		8 x6 = 48 7 x7 = 49
0 x7 = 0		3 x3 = 9	5 x9 = 45

Copyright © 1991 American Education Publishing Co.

Multiplication: Tens, Hundreds, And Thousands

Multiplication is a short way to find the sum of adding the same number a certain amount of times, such as 7 x 4 = 28 instead of 7 + 7 + 7 + 7 = 28.

Directions: Study the examples.

Examples:

When multiplying a number by 10, the answer is the number with a zero. It is like counting by 10s.

10 x1 = 10	10 x2 = 20	10 x3 = 30	10 x4 = 40	10 x5 = 50	10 x6 = 60

When multiplying a number by 100, the answer is the number with two zeroes.
When multiplying a number by 1000, the answer is the number with three zeroes.

100 x1 = 100	100 x2 = 200	100 x3 = 300	1000 x1 = 1000	1000 x2 = 2000	1000 x3 = 3000

Such basic facts help us multiply.

4 x2 = 8	400 x2 = 800	8 x3 = 24	800 x3 = 2400	7 x5 = 35	700 x5 = 3,500

Directions: Multiply.

10 x3 = 30	60 x5 = 300	400 x5 = 2000	700 x8 = 5600	50 x7 = 350
80 x9 = 720	4000 x2 = 8000	6000 x4 = 24,000	300 x9 = 2700	700 x6 = 4200

3 x 800 = 2400   9 x 2000 = 18,000   7 x 90 = 630

Copyright © 1991 American Education Publishing Co.

Multiplication: One-Digit Number x Two-Digit Number

Multiplication is a short way to find the sum of adding the same number a certain amount of times, such as 7 x 4 = 28 instead of 7 + 7 + 7 + 7 = 28.

Directions: Study the example. Follow the steps to multiplying by regrouping tens.

Example:

Step 1. Multiply ones. Regroup.
54 x7 = 8

Step 2. Multiply Tens. Add 2 tens.
54 x7 = 378

27 x3 = 81	63 x4 = 252	52 x5 = 260	91 x9 = 819	45 x7 = 315
75 x2 = 150	64 x5 = 320	76 x3 = 228	93 x6 = 558	87 x4 = 348
66 x7 = 462	38 x2 = 76	47 x8 = 376	64 x9 = 576	51 x8 = 408
99 x3 = 297	13 x7 = 91	32 x4 = 128	25 x8 = 200	15 x7 = 105

The chickens on the Smith farm produce 48 dozen eggs each day. How many dozen eggs do they produce in 7 days?   336

Copyright © 1991 American Education Publishing Co.

Multiplication: Two-Digit Number x Two-Digit Number

Multiplication is a short way to find the sum of adding the same number a certain amount of times, such as 7 x 4 = 28 instead of 7 + 7 + 7 + 7 = 28.

Directions: Study the examples. Follow the steps to multiply by regrouping.

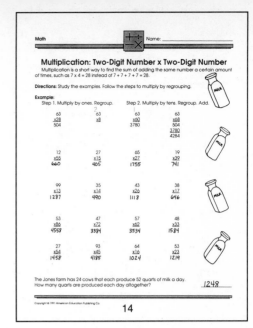

Example:

Step 1. Multiply by ones. Regroup.
63 x28 = 504   63 x8

Step 2. Multiply by tens. Regroup. Add.
63 x50 = 3780   63 x68 = 504 / 3780 / 4284

12 x55 = 660	27 x15 = 405	65 x27 = 1755	19 x39 = 741
99 x13 = 1287	35 x14 = 490	43 x26 = 1118	38 x17 = 646
53 x86 = 4558	47 x72 = 3384	57 x62 = 3534	48 x33 = 1584
27 x54 = 1458	93 x45 = 4185	64 x16 = 1024	53 x23 = 1219

The Jones farm has 24 cows that each produce 52 quarts of milk a day. How many quarts are produced each day altogether?   1248

Copyright © 1991 American Education Publishing Co.

Multiplication: Two-Digit Number x Three-Digit Number

Multiplication is a short way to find the sum of adding the same number a certain amount of times, such as 7 x 4 = 28 instead of 7 + 7 + 7 + 7 = 28.

Directions: Study the example. Follow the steps to multiply.

Example:

Step 1. Multiply by ones. Regroup.
287 x43 = 861   287 x3 = 861

Step 2. Multiply by tens. Regroup. Add.
287 x40 = 11,480   287 x43 = 861 / 11,480 / 12,341

261 x36 = 9,396	434 x48 = 20,832	357 x75 = 26,775	614 x59 = 36,226	368 x98 = 36,064
231 x46 = 10,626	754 x65 = 49,010	549 x89 = 48,861	372 x94 = 34,968	458 x85 = 38,930

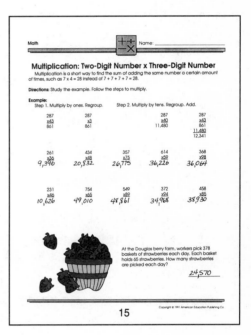

At the Douglas berry farm, workers pick 378 baskets of strawberries each day. Each basket holds 65 strawberries. How many strawberries are picked each day?   24,570

Copyright © 1991 American Education Publishing Co.

Multiplication: Three-Digit Number x Three-Digit Number

Multiplication is a short way to find the sum of adding the same number a certain amount of times, such as 7 x 4 = 28 instead of 7 + 7 + 7 + 7 = 28.

Directions: Multiply. Regroup when needed.

Hint: When multiplying by the tens, start writing the number in the tens place. When multiplying by the hundreds, start in the hundreds place.

Example:

563 x248 = 4504 / 2252 / 1126 / 139624

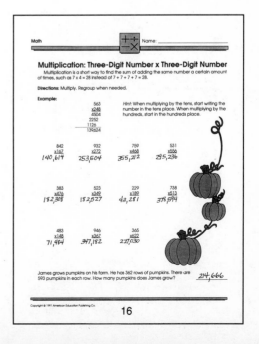

842 x167 = 140,614	932 x272 = 253,604	759 x468 = 355,212	531 x556 = 295,236
383 x476 = 182,308	523 x349 = 182,527	229 x189 = 43,281	738 x513 = 378,594
483 x148 = 71,484	946 x367 = 347,182	365 x622 = 227,030	

James grows pumpkins on his farm. He has 362 rows of pumpkins. There are 593 pumpkins in each row. How many pumpkins does James grow?   214,666

Copyright © 1991 American Education Publishing Co.

Multiplication

Multiplication is a short way to find the sum of adding the same number a certain amount of times, such as 7 x 4 = 28 instead of 7 + 7 + 7 + 7 = 28.

Directions: Multiply. Use your answers to follow the code to color the quilt.

| 70,725 — red | 448 — white | 34,088 — blue |
| 667 — green | 249,738 — orange | 221,446 — yellow |

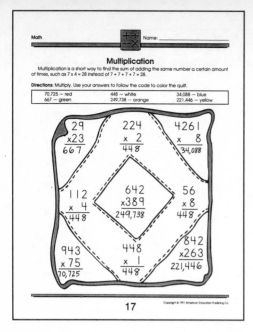

Copyright © 1991 American Education Publishing Co.

Division: Checking The Answer

Division is a way to find out how many times one number is contained in another number. To check a division problem, multiply the quotient by the divisor. Add the remainder. The answer will be the dividend.

Directions: Study the example. Divide to work the problems. Draw a line from the division problem to the correct checking problem.

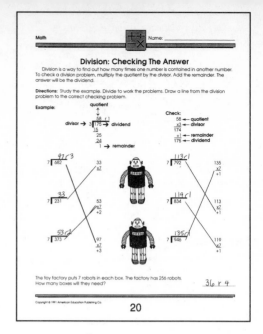

The toy factory puts 7 robots in each box. The factory has 256 robots. How many boxes will they need? 36 r 4

Copyright © 1991 American Education Publishing Co.

Review

Directions: Multiply. Work the problem in the box. Color the ribbons blue if the answer is correct.

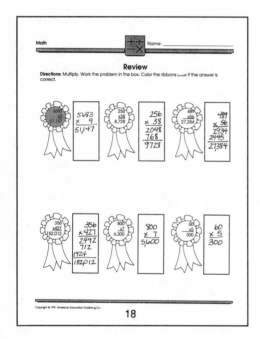

Copyright © 1991 American Education Publishing Co.

Division: 1-Digit Divisor

Division is a way to find out how many times one number is contained in another number.

Directions: Work the problems on another sheet of paper. Use the code to color the picture.

Color these answers:

5⟌895 (A=179)	6⟌493 (A=82 r1)	6⟌940 (A=156 r4)	4⟌647 (A=161 r3) orange
4⟌672 (A=168)	6⟌696 (A=116)	5⟌745 (A=149 r4)	8⟌628 (A=78 r4) blue
3⟌814 (A=271 r1)	7⟌490 (A=70)	5⟌398 (A=79 r3)	2⟌571 (A=285 r1) black

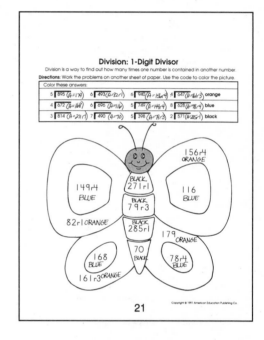

Copyright © 1991 American Education Publishing Co.

Division

Division is a way to find out how many times one number is contained in another number. For example, 28 ÷ 7 = 4 means that there are four groups of seven in 28.

Directions: Study the example. Then divide to solve the problems. Remember that the remainder must be smaller than the divisor.

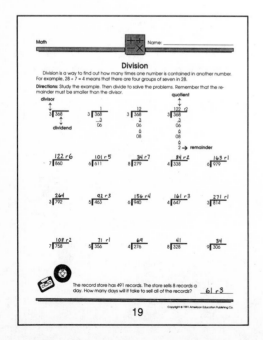

The record store has 491 records. The store sells 8 records a day. How many days will it take to sell all of the records? 61 r 3

Copyright © 1991 American Education Publishing Co.

Division: 2-Digit Divisor

Division is a way to find out how many times one number is contained in another number.

Directions: Study the example. Divide. Remember to check your answer by multiplying it by the divisor and adding the remainder.

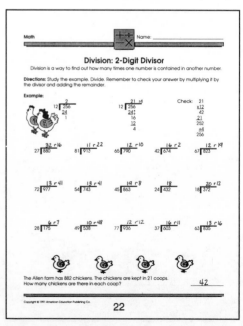

The Allen farm has 882 chickens. The chickens are kept in 21 coops. How many chickens are there in each coop? 42

Copyright © 1991 American Education Publishing Co.

Division: Checking The Answer

Division is a way to find out how many times one number is contained in another number.

Directions: Divide, then check your answers.

Example:

```
    182  r1      Check:     182
4 | 729                      x4
    4                       728
    32                      + 1
    32                      729
     9
     8
     1
```

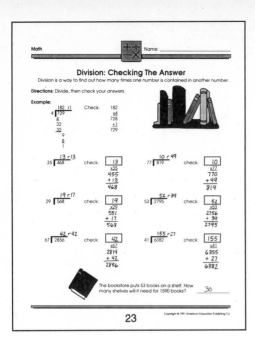

```
     13 r13    check:  | 13 |
35 | 468                 x35
                        455
                       + 13
                        468
```

```
     10 r 49   check:  | 10 |
77 | 819                 x77
                        770
                       + 49
                        819
```

```
     19 r17    check:  | 19 |
29 | 568                 x29
                        551
                       + 17
                        568
```

```
     52 r39    check:  | 52 |
53 | 2795               x53
                       2756
                       + 39
                       2795
```

```
     42 r42    check:  | 42 |
67 | 2856               x67
                       2814
                       + 42
                       2856
```

```
     155 r27   check:  | 155 |
41 | 6382               x41
                       6355
                       + 27
                       6382
```

The bookstore puts 53 books on a shelf. How many shelves will it need for 1590 books?　　30

23

Review

Directions: Divide.

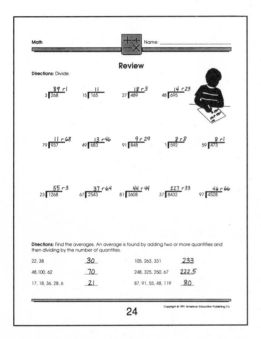

```
    89 r1         11          18 r3       14 r23
3 | 268      15 | 165     27 | 489     48 | 695
```

```
    11 r68        13 r46       9 r29        8 r8        8 r1
79 | 937     49 | 683     91 | 848     1 | 592     59 | 473
```

```
    55 r3         37 r64       44 r 44      227 r33      46 r 66
23 | 1268    67 | 2543    81 | 3608    37 | 8432    97 | 4528
```

Directions: Find the averages. An average is found by adding two or more quantities and then dividing by the number of quantities.

22, 38	30	105, 263, 331	233
48, 100, 62	70	248, 325, 250, 67	222.5
17, 18, 36, 28, 6	21	87, 91, 55, 48, 119	80

24

Fraction: Addition

A fraction is a number that names part of a whole, such as 1/2 or 1/3. The denominator is the bottom number in a fraction; the numerator is the top number.
When adding fractions with the same denominator, the denominator stays the same. Add only the numerators.

Example:
numerator → $\frac{1}{8}$ + $\frac{2}{8}$ = $\frac{3}{8}$ ← denominator

Directions: Study the example. Add the fractions. The first one is done for you.

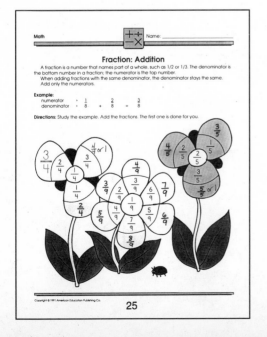

25

Fractions: Subtraction

A fraction is a number that names part of a whole, such as 1/2 or 1/3. The denominator is the bottom number in a fraction; the numerator is the top number.
When subtracting fractions with the same denominator, the denominator stays the same. Subtract only the numerators.

Directions: Solve the problems below, working from left to right across each row. As you find each answer, copy the letter from the code box into the numbered blanks. The first one is done for you. The answer will tell the name of a famous American.

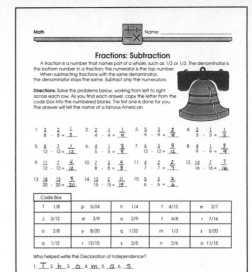

1. $\frac{3}{8} - \frac{2}{8} = \frac{1}{8}$　　2. $\frac{3}{4} - \frac{2}{4} = \frac{1}{4}$　　3. $\frac{5}{9} - \frac{3}{9} = \frac{2}{9}$　　4. $\frac{2}{3} - \frac{1}{3} = \frac{1}{3}$

5. $\frac{8}{12} - \frac{7}{12} = \frac{1}{12}$　　6. $\frac{4}{12} - \frac{1}{12} = \frac{3}{12}$　　7. $\frac{6}{12} - \frac{3}{12} = \frac{3}{12}$　　8. $\frac{4}{12} - \frac{1}{12} = \frac{3}{12}$

9. $\frac{11}{12} - \frac{7}{12} = \frac{4}{12}$　　10. $\frac{7}{8} - \frac{5}{8} = \frac{2}{8}$　　11. $\frac{4}{7} - \frac{2}{7} = \frac{2}{7}$　　12. $\frac{14}{16} - \frac{7}{16} = \frac{7}{16}$

13. $\frac{18}{20} - \frac{13}{20} = \frac{5}{20}$　　14. $\frac{13}{15} - \frac{2}{15} = \frac{11}{15}$　　15. $\frac{5}{6} - \frac{3}{6} = \frac{2}{6}$

Code Box									
T	1/8	p	5/24	h	1/4	f	4/12	e	2/7
J	3/12	e	3/9	o	2/9	f	4/8	r	7/16
o	2/8	y	8/20	q	1/32	m	1/3	s	5/20
a	1/12	r	12/15	s	3/5	n	2/6	o	11/15

Who helped write the Declaration of Independence?

1. T 2. h 3. o 4. m 5. a 6. s
7. J 8. e 9. f 10. f 11. e 12. r 13. s 14. o 15. n

26

Fractions: Adding Mixed Numerals

A mixed numeral is a number written as a whole number and a fraction, such as 6 5/8.

Directions: Add the number in the center to the numbers in the rings.

Example:

```
   9  1/3              2  3/6
  +3                  +1
  12  2/3              4  4/6
```

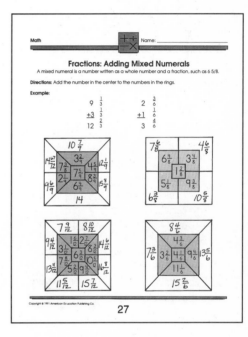

27

Fractions: Subtracting Mixed Numerals

A mixed numeral is a number written as a whole number and a fraction, such as 6 5/8.

Directions: Solve the problems. The first one is done for you.

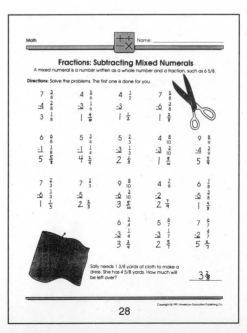

```
  7  3/8      4  5/6      4  1/2      7  5/8
 -4  2/8     -3  1/6     -3         -6  1/8
  3  1/8      1  4/6      1  1/2      1  4/8
```

```
  6  6/8      5  3/4      5  2/4      4  8/10     9  8/9
 -1  1/8     -1  1/4     -3  1/4     -3  2/10    -4  4/9
  5  5/8      4  2/4      2  1/4      1  6/10     5  4/9
```

```
  7  2/8      7  3/4      9  8/10     4  7/9      6  7/8
 -6  1/8     -5         -6  3/10    -2  1/9     -5  4/8
  1  1/8      2  3/4      3  5/10     2  6/9      1  3/8
```

```
              6  3/4      5  6/7      7  6/9
             -3  1/4     -3  1/7     -2  1/9
              3  2/4      2  5/7      5  5/9
```

Sally needs 1 3/8 yards of cloth to make a dress. She has 4 5/8 yards. How much will be left over?　　3 2/8

28

Fractions: Equivalent

Equivalent fractions name the same number, such as 1/2 and 2/4.

Directions: Study the example. Draw a line between the equivalent fractions in each row.

Example:
Equivalent fractions are equal to each other.
$$1/2 = 2/4 = 4/8$$

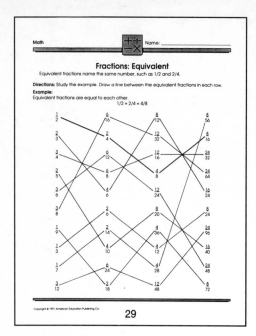

Fractions: Reducing

Reducing a fraction means to find the greatest common factor and divide.

Directions: Reduce each fraction. Circle the answer.

Example:　$\frac{5}{15} = 3$　factors of 5:　1, 5　　　$5 \div 5 = 1$
　　　　　　　　　　　factors of 15:　1, 3, 5, 15　　$15 \div 5 = 3$

$\frac{2}{4} = \frac{1}{2}, \frac{1}{6}, \frac{1}{8}$　　　$\frac{3}{9} = \frac{1}{6}, \frac{1}{3}, 3$　　　$\frac{5}{10} = \frac{1}{5}, \frac{1}{2}, \frac{5}{6}$

$\frac{4}{12} = \frac{1}{4}, \frac{1}{3}, 3$　　　$\frac{10}{15} = \frac{2}{3}, \frac{2}{5}, \frac{2}{2}$　　　$\frac{12}{14} = \frac{1}{8}, \frac{6}{7}, \frac{3}{5}$

$\frac{3}{24} = \frac{3}{12}, \frac{3}{6}, \frac{1}{8}$　　　$\frac{1}{11} = \frac{1}{11}, \frac{3}{5}, \frac{1}{4}$　　　$\frac{11}{22} = \frac{1}{12}, \frac{1}{2}, \frac{3}{5}$

Directions: Find the way home. Color the boxes with fractions equivalent to 1/8 and 1/3.

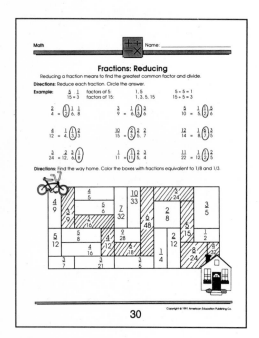

Fractions: Mixed Numerals

A mixed numeral is a number written as a whole number and a fraction, such as 6 5/8.

Directions: Change each fraction to a mixed numeral. Make the mixed numerals into fractions.

Example:
To change a fraction into a mixed numeral, divide the denominator (bottom number) into the numerator (top number). Put the remainder over the denominator.
To change a mixed numeral into a fraction, multiply the denominator by the whole number, add the numerator, and place it on top of the denominator.

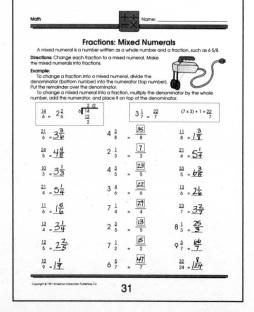

Review

Directions: Add or subtract the fractions and mixed numerals.

$\frac{3}{8} - \frac{1}{8} = \frac{2}{8}$　　$\frac{3}{4} - \frac{1}{4} = \frac{4}{}$　　$\frac{3}{5} + \frac{1}{5} = \frac{4}{5}$　　$\frac{3}{12} + \frac{1}{12} = \frac{7}{12}$　　$\frac{3}{9} + \frac{1}{9} = \frac{4}{9}$

$3\frac{1}{8}$	$4\frac{5}{6}$	$7\frac{1}{11}$	$8\frac{3}{9}$	$4\frac{7}{8}$
$+1\frac{3}{8}$	$-3\frac{1}{6}$	$+3\frac{3}{11}$	$+2\frac{5}{9}$	$-2\frac{5}{8}$
$4\frac{4}{8}$	$1\frac{4}{6}$	$10\frac{3}{11}$	$10\frac{8}{9}$	$2\frac{2}{8}$

Directions: Reduce the fractions. Circle the answers.

$\frac{3}{6} = \frac{1}{7}, \frac{1}{2}, \frac{1}{4}$	$\frac{2}{8} = \frac{1}{3}, \frac{1}{4}, \frac{1}{16}$	$\frac{4}{6} = \frac{1}{4}, \frac{2}{3}, \frac{3}{9}$
$\frac{4}{20} = \frac{1}{4}, \frac{3}{5}, \frac{1}{5}$	$\frac{7}{21} = \frac{1}{7}, \frac{1}{3}, \frac{1}{5}$	$\frac{9}{12} = \frac{3}{5}, \frac{1}{8}, \frac{3}{4}$

Directions: Reduce the fractions.

$\frac{6}{24} = \frac{1}{4}$　　　$\frac{8}{32} = \frac{1}{4}$　　　$\frac{2}{4} = \frac{1}{2}$

$\frac{3}{15} = \frac{1}{5}$　　　$\frac{6}{12} = \frac{1}{6}$　　　$\frac{3}{9} = \frac{1}{3}$

Directions: Change the mixed numerals to fractions and the fractions to mixed numerals.

$3\frac{1}{3} = \frac{10}{3}$　　$\frac{14}{4} = 3\frac{1}{2}$　　$\frac{26}{6} = 4\frac{1}{3}$　　$3\frac{7}{12} = \frac{43}{12}$　　$\frac{22}{7} = 3\frac{1}{7}$

NOTES